SILVERSIDE SHRIMP FARMING

Building A Profitable Shrimp Aquaculture Business

Explore The Lucrative World Of Silverside Shrimp Farming For Gourmet Seafood Markets

Dr. Fabian Felicity

Table of Contents

CHAPTER ONE

Introduction

Shrimp farming, namely Silverside Shrimp Farming, has become a successful undertaking in the aquaculture business. Global demand for shrimp is increasing, driven by a growing consumer appetite for seafood and improved awareness of the health benefits of shrimp consumption.

As a result, companies and investors are exploring shrimp farming opportunities, with the Silverside Shrimp being a particularly attractive species. This article delves into the complexity of Silverside Shrimp Farming, emphasizing the market potential and the significance of choosing the right location for a successful shrimp farm.

Understand Silverside Shrimp Farming

Silverside Shrimp, also known as Litopenaeus Schmitt, is a popular shrimp species because of its exquisite meat and rapid growth as compared to other shrimp species.

Silverside Shrimp, which is native to the Atlantic and Gulf of Mexico, has emerged as a vital participant in shrimp aquaculture due to its adaptability in farming conditions and capacity to thrive in a variety of environments.

Silverside Shrimp Farming is strongly dependent on the shrimp's life cycle. Typically, the approach begins with the selection of high-quality brood stock.

These are adult shrimp selected for their genetic characteristics, ensuring that beneficial traits are handed down to offspring. The broodstock is well-cared for to achieve optimal reproduction.

The next phase involves hatcheries, where shrimp larvae are reared until they are ready for transfer to grow-out ponds. During this time, the larvae are carefully cared for in a controlled environment that mimics their natural habitat. Feed, water quality, and temperature are carefully monitored to guarantee proper growth.

When the larvae mature into post-larvae, they are moved to grow-out ponds to finish their development. Grow-out pond management is crucial, and it covers factors like water quality, stocking density, and disease control.

Silverside Shrimp farming often employs novel technology to monitor and change these variables, resulting in a sustainable and effective farming approach.

Market Opportunities In The Shrimp Aquaculture Industry

The global shrimp aquaculture business has grown dramatically in recent years, due to increased demand and a shift toward more environmentally friendly fishing practices. Silverside Shrimp, with its favorable qualities, provides an ideal opportunity for companies looking to capitalize on this trend.

One of the key factors driving market growth in the shrimp aquaculture industry is rising consumer awareness of the health benefits of shrimp consumption. Shrimp are a low-calorie protein rich in omega-3 fatty acids, vitamins, and minerals. As health-conscious consumers seek nutritious alternatives, demand for high-quality shrimp products, such as Silverside Shrimp, is expected to grow.

Furthermore, the globalization of food supply chains has opened up new markets for shrimp goods. Emerging economies with a growing middle class are seeing an increase in disposable income, driving rising demand for seafood.

Silverside Shrimp, with its culinary versatility, is well-positioned to fulfill a diverse spectrum of client preferences.

Another element that influences market prospects is the sustainability of shrimp farming. Consumers are increasingly seeking sustainable and environmentally friendly food sources. When done appropriately, Silverside Shrimp farming may correlate to these preferences, resulting in a positive market image.

Choosing The Best Location For Your Shrimp Farm:

The proper location is a significant determinant of success in Silverside Shrimp Farming. Several factors must be addressed to provide the optimal conditions for shrimp growth and development, as well as the overall sustainability of the farming enterprise.

Shrimp farming needs high-quality water since these crustaceans are very sensitive to variations in salinity, temperature, and oxygen levels. As a consequence, access to a secure and reliable water supply is essential. Additionally, comprehensive water testing and analysis should be undertaken to verify its suitability for shrimp farming.

Climate has a significant influence on shrimp farming performance. Silverside Shrimp flourishes in tropical and subtropical climates, particularly in places with constant temperatures throughout the year. Extreme temperature fluctuations may harm shrimp and limit their growth. Furthermore, selecting a location with a low risk of natural disasters such as hurricanes or typhoons is crucial for preventing damage to shrimp ponds and infrastructure.

When deciding where to set up a shrimp farm, infrastructure and accessibility are crucial considerations. The farm requires enough infrastructure, such as roads, electricity, and water supply, to work properly. The closeness of transportation hubs facilitates the conveyance of harvested shrimp to markets, reducing logistical issues and expenses.

Local laws and community support are significant considerations. Establishing a shrimp farm requires adhering to environmental guidelines and obtaining the necessary permits. Engaging with the local community and gaining their support may help the firm stay viable in the long term. Addressing potential environmental hazards and using ethical agriculture techniques might help you create solid relationships with local stakeholders.

To conclude, Silverside Shrimp Farming is a promising aquaculture venture, fueled by expanding global demand for shrimp and a growing emphasis on sustainable fishing practices. Success in this approach is contingent on a detailed understanding of the shrimp's life cycle, market dynamics, and careful selection of an appropriate location. Entrepreneurs and investors who can navigate these hurdles will be well-positioned to capitalize on the shrimp farming industry's growing potential.

CHAPTER TWO
Designing And Building
Shrimp Ponds

Shrimp farming has expanded into a big industry, offering a lucrative source of income for many aquaculturists. One of the most significant parts of successful shrimp farming is the careful design and construction of shrimp ponds. These ponds offer an environment for shrimp growth and development, affecting the farm's overall production.

Factors Influencing Pond Design

Before commencing the construction of shrimp ponds, many issues must be solved. The topography of the location, soil quality, and water supply all play vital roles in determining the sustainability of pond development.

Ponds should be built in areas with a moderate slope to improve water circulation and limit the chance of waterlogging.

The size and architecture of the ponds are major factors. Larger ponds are often less expensive, allowing for economies of scale in shrimp production. Rectangular or square forms are often used because they increase space and provide constant water distribution.

Pond Liners And Construction Materials

Choosing the right construction materials is crucial to the longevity and efficiency of shrimp ponds. Pond liners, composed of high-density polyethylene (HDPE) or clay, help to prevent water seepage and maintain the water level where it should be. HDPE liners are utilized for their durability and ease of installation.

Pond embankments should be properly constructed to provide stability and resistance to erosion.

Adding aeration equipment, such as paddlewheels or diffused air systems, raises oxygen levels and promotes perfect water conditions for shrimp growth.

Water Quality Control For Silverside Shrimp Farming

The success of a shrimp farm is heavily reliant on maintaining good water quality. Poor water quality may lead to stress, disease outbreaks, and impaired shrimp growth. Effective water quality management is so crucial in silverside shrimp farming.

Monitoring and Control Parameters

Regular monitoring of water parameters is critical for spotting potential issues early on. Temperature, salinity, pH, dissolved oxygen, and ammonia levels must be measured regularly. Advanced monitoring systems, such as sensors and automated control systems, may help to ensure real-time adjustments to maintain ideal conditions.

Aeration And Water Exchange

Aeration is critical for maintaining dissolved oxygen levels in shrimp ponds. Aeration not only increases oxygenation, but it also helps to distribute heat evenly, eliminating thermal stratification. Proper water exchange is essential for both garbage removal and maintaining appropriate salinity levels.

Selected And Sourced Shrimp Broodstock

The selection of high-quality broodstock is critical to the success of any shrimp farming enterprise. Broodstock are adult shrimp selected for breeding, and their genetic quality has a significant influence on the offspring's characteristics.

Proper broodstock selection and procurement are key aspects of running a successful shrimp farm.

CHAPTER THREE

Genetic Factors

Choosing broodstock with suitable genetic characteristics is crucial. Genetic diversity, sickness resistance, and development ability are all significant factors to consider. Working with reputable hatcheries that prioritize genetic selection programs may provide access to high-quality broodstock.

Health Screening and Quarantine

Before importing broodstock to the farm, thorough health screening and quarantine processes should be carried out. This helps to prevent the spread of diseases that might affect the whole shrimp population.

Regular health checks and immunizations, where required, help to build a sound biosecurity program.

Shrimp Hatchery Operations: From Larva To Postlarvae

The shift from shrimp larvae to postlarvae is an essential step in shrimp farming. Successful hatchery operations are crucial for ensuring a consistent and healthy supply of shrimp for grow-out ponds.

This intricate technique requires meticulous attention to detail as well as adherence to the best standards.

Broodstock Maturation And Spawning

Hatchery operations begin with the maturation of broodstock and the start of spawning. Creating ideal conditions for broodstock, such as maintaining proper water parameters and providing a nutritious diet, is crucial for successful maturation. Spawning procedures vary, although natural and induced approaches are common.

Larval Rearing And Feeding

Once hatched, shrimp larvae need special care. Larval rearing rooms with controlled environmental conditions are used to enhance growth. Proper nutrition is crucial during this stage, and live feeds such as microalgae and Artemia are often employed. Feeding regimens must be strictly followed to provide optimal growth and survival rates.

Post-Larvae Production And Nursery Management

The last stage of hatchery operations is the production of postlarvae, or young shrimp suitable for stocking in grow-out ponds. During this time, proper nursery care is critical for gradually acclimating postlarvae to pond conditions.

CHAPTER FOUR

Disease Management and Biosecurity

To prevent disease spread, strong biosecurity measures must be implemented throughout hatchery operations. Regular health checks, quarantine regulations, and the use of probiotics all help to ensure successful disease management.

Water quality in the hatchery must be constantly monitored to provide a suitable environment for postlarvae development.

To recap, the success of a shrimp farming business is contingent on paying close attention to each step of the production process. From shrimp pond design to hatchery operations, a comprehensive approach is necessary to ensure long-term and profitable shrimp farming.

Feeding Strategy For Healthy Shrimp Growth

In the ever-expanding world of aquaculture, optimizing feeding procedures is vital to ensuring shrimp populations' health and survival. Shrimp farming has evolved dramatically, and the industry's success is largely based on the implementation of effective feeding practices. Several essential components are crucial in promoting healthy shrimp growth, ranging from selecting appropriate feed formulations to executing feeding regimens adapted to the unique needs of each shrimp species.

Food Formulations

The composition of shrimp feed is the foundation for any successful feeding strategy. A well-balanced diet is essential to meet the nutritional demands of shrimp at various life stages. Protein, a key component of shrimp growth, should be well-balanced in the diet mix. High-quality protein sources, such as fishmeal and soybean meal, are often used to provide appropriate protein content.

Furthermore, essential vitamins and minerals must be incorporated into the shrimp diet to ensure overall health. Micronutrients including vitamins C and E aid in boosting the shrimp's immune system and stress tolerance. Feeds with the right nutrient balance ensure that shrimp get a nutritionally balanced meal, promoting healthy growth and development.

Feeding Frequency And Timing

In shrimp aquaculture, feeding frequency and timing are as critical as feed content. Shrimp are well-known for their voracious eating habits, but overfeeding may have harmful repercussions for both water quality and health. It is vital to devise a feeding strategy that is compatible with the shrimp's natural activity and metabolism.

Feeding frequency varies according to shrimp species, pond conditions, and environmental factors. Some farmers provide several microfeedings throughout the day to mimic the shrimp's natural grazing behavior. Others may like fewer, larger meals.

Observing shrimp behavior and monitoring water quality measurements may help determine the ideal feeding strategy for maximal growth and health.

CHAPTER FIVE
Disease Prevention And Management In Shrimp Aquaculture

The rapidly expanding shrimp aquaculture industry is constantly threatened by a variety of diseases, which may have a significant impact on output. Disease prevention and control techniques are crucial for reducing the risks presented by bacterial, viral, and parasite infections, which may spread rapidly in densely populated shrimp farms.

Biosecurity Measures

Strict biosecurity measures are the first line of defense against illnesses in shrimp production. This includes controlling and monitoring the flow of people, equipment, and water in and out of the farm. Furthermore, frequent health checks of incoming shrimp stock help spot potential disease carriers before they spread across the colony.

Farmers also use safe water treatment techniques, such as the use of probiotics and beneficial bacteria. These measures maintain a healthy pond ecosystem, reducing the likelihood of disease outbreaks. Adequate pond preparation and disinfection before stocking new shrimp batches are critical components of a comprehensive biosecurity plan.

Vaccines And Immunostimulants

In recent years, improvements in shrimp sickness treatment have led to the development of vaccines and immunostimulants. Shrimp vaccination programs aim to strengthen their immune response to certain infections, hence reducing their susceptibility to sickness. Immunostimulants, on the other hand, enhance the whole immune system, providing a proactive defense mechanism against a variety of diseases.

Integrating these novel approaches into shrimp farming operations requires a thorough analysis of the target illnesses as well as the shrimp species' specific immunity needs. As research in shrimp immunology advances, the aquaculture industry will benefit from more effective and sustainable disease prevention strategies.

Harvesting Techniques For Silverside Shrimp

Shrimp aquaculture's success extends beyond the growth phase; proper harvesting processes are equally essential for assuring the quality and market value of the final product.

Silverside shrimp, a popular aquaculture species, need careful handling and precision throughout the harvesting process to retain their delicate flavor and texture.

Selecting The Right Harvesting Tools

To minimize stress and harm to the delicate crustaceans, silverside shrimp must be collected using special tools. Shrimp nets and seines are typical methods for catching shrimp without hurting them. The mesh size of this equipment has been carefully selected to allow smaller shrimp to escape, ensuring that only marketable-sized shrimp are caught.

Furthermore, gentle harvesting equipment like vacuum pumps has gained popularity for lowering stress on shrimp during the catch. These novel processes contribute to increased survival rates and improved product quality, meeting the stringent requirements of gourmet seafood markets.

Harvesting At Optimum Size And Maturity

Timing is critical in shrimp harvesting to ensure that the product meets market expectations. Harvesting shrimp at the optimal size and maturity is essential for getting the desired taste and texture. Oversized or underdeveloped shrimp may not fetch the premium costs associated with gourmet seafood establishments.

Farmers closely monitor the growth of their shrimp populations and use size-selective plucking to create batches that meet market requirements. This precision in harvesting improves not only the economic viability of shrimp farming but also the enjoyment of consumers searching for high-quality seafood.

Process And Packaging For The Gourmet Seafood Market

Shrimp's journey from the aquaculture farm to the gourmet seafood market requires rigorous processing and packing to ensure the freshness and quality that discerning purchasers demand.

CHAPTER SIX

Chilling And Freezing Technology

The procedure of maintaining shrimp quality begins soon after harvest. Chilling and freezing techniques are used to keep shrimp fresh and inhibit the growth of spoilage-causing microorganisms.

Rapid chilling processes, such as blast freezing, reduce the development of ice crystals, which may damage shrimp texture.

Market demand and practical restrictions determine whether fresh or frozen shrimp are used. Some gourmet seafood restaurants stress the ease of frozen goods, while others highlight the better quality of fresh, never-frozen shrimp.

Sustainable Packaging Practices

As consumer awareness of environmental issues grows, the seafood industry faces greater pressure to use sustainable packaging practices. Gourmet seafood markets often choose suppliers that concentrate on eco-friendly packaging materials with little environmental impact.

Biodegradable and recyclable packaging, such as biodegradable trays and recycled cardboard, are more popular. These choices not only reflect client preferences but also strengthen the industry's commitment to environmental responsibility. Furthermore, appropriate and safe packing is essential to prevent damage during transportation and storage, ensuring that the shrimp reaches the buyer in excellent condition.

Finally, the success of shrimp aquaculture necessitates a multifaceted approach, spanning from creating efficient feeding systems for healthy growth to implementing effective disease prevention and control protocols.

Shrimp meet the high standards of gourmet seafood markets owing to sophisticated harvesting techniques, as well as rigorous processing and packaging. Balancing these elements fosters a thriving shrimp aquaculture industry that meets the demands of both producers and consumers.

Market Your Silverside Shrimp Products

In the competitive world of seafood, marketing is vital to the success of any shrimp farming enterprise, especially when promoting a distinctive product like Silverside shrimp. Understanding and using Silverside Shrimp's distinct selling points is critical for establishing a successful marketing strategy.

Understanding the Market

Before starting marketing activities, it is necessary to do significant market research. It is vital to understand the preferences of your target audience as well as the current market trends. uncover key competitors and analyze their techniques to uncover opportunities for differentiation.

Branding And Positioning

Establishing a strong brand identity is critical. Create a brand that emphasizes the exceptional quality and unique characteristics of Silverside Shrimp.

Create a unique market positioning that sets Silverside Shrimp apart from other shrimp offers. Highlight the aspects that distinguish your goods, such as taste, texture, or sustainable agriculture processes.

Online Presence

In the digital age, having a great online presence is not a choice. Create a user-friendly website offering Silverside Shrimp products, replete with high-quality images and detailed product information.

Use social media to interact with new customers, promote recipes, and highlight the sustainable practices that underpin Silverside Shrimp farming.

CHAPTER SEVEN

Targeted Marketing Campaigns

Create personalized marketing tactics for certain customer groups. Highlight Silverside Shrimp's health benefits, such as high protein content and omega-3 fatty acids.

Collaborate with chefs or nutritionists to create content that educates people on the nutritional value and versatility of Silverside Shrimp in diverse cuisines.

Distribution Channels

Establish efficient distribution channels to ensure that Silverside Shrimp products are widely available to consumers. Collaborate with supermarkets, restaurants, and seafood markets to expand your reach. Consider e-commerce platforms to reach a broader audience and provide more convenient online purchasing options.

Financial Planning And Management For Shrimp Farming

A shrimp farming company's long-term viability is dependent on strong financial planning and management. From initial investment to ongoing running expenses, effective financial management fosters profitability and resilience in the face of economic downturns.

Start-Up Costs

Start by doing a detailed evaluation of beginning expenditures. This includes shrimp farm infrastructure, equipment, hatchery fees, and other initial costs. Create a clear budget to calculate the amount of money required, and consider funding options such as loans or grants.

Operating Expenses

Identify and categorize ongoing operational expenses. Possible examples include feed expenditures, water and electricity charges, labor costs, and maintenance fees. Implement cost-cutting techniques while maintaining the quality of Silverside Shrimp production.

Revenue Projections

Make accurate revenue projections based on market demand, pricing strategies, and production capacity. Consider a variety of scenarios and market volatility while creating a flexible financial model. As the company expands, review projections regularly and adjust them as needed.

Risk Management

Diseases, environmental concerns, and market volatility are all threats to shrimp farming. Implement risk management strategies to minimize potential losses. This may involve investing in current monitoring systems, insurance coverage, and increasing product offerings.

Financial Monitoring And Reporting

Set up excellent financial monitoring systems to track income, expenses, and profits. Regularly review financial data to assess company health and identify areas for development. Use accounting software to streamline financial processes and ensure regulatory compliance.

Sustainable Shrimp Aquaculture

Sustainability is an essential aspect of modern aquaculture, and incorporating sustainable practices into Silverside Shrimp production is not only environmentally responsible but also boosts the product's economic desirability.

Responsible Farming Practices

Use acceptable and ethical farming techniques that prioritize the health of the shrimp and the surrounding ecosystem. Install water recirculation systems to minimize water use and waste. To prevent environmental harm, ensure that rubbish is appropriately handled.

Certifications And Labels

Obtain certifications and labels demonstrating the sustainability and quality of Silverside Shrimp products. Certifications from reputable organizations, such as the Aquaculture Stewardship Council (ASC) or the Global Aquaculture Alliance (GAA), may increase consumer confidence and open up new markets.

Community Engagement

Engage with local communities to build solid connections. Implement social responsibility programs that benefit the community and promote sustainable behavior. Transparent information regarding the shrimp farm's environmental practices may boost consumer trust.

Research And Innovation

Invest in research and innovation to continually improve your sustainable practices. Investigate alternative feed sources, energy-efficient solutions, and novel farming practices to reduce the environmental impact of Silverside Shrimp production.

Educating Customers

Customers should be educated on the importance of buying sustainable seafood and the advantages of supporting products that are produced responsibly. Create marketing campaigns that highlight Silverside Shrimp's commitment to sustainability and position it as a key selling point.

Conclusion

To conclude, understanding the sector, developing a strong brand, and using online channels are critical for effectively marketing Silverside Shrimp products.

Financial planning and management are crucial to the long-term viability of the shrimp farming industry and include careful consideration of start-up costs, operational expenses, revenue projections, and risk management.

Integrating sustainable processes is not only an ethical option, but it also improves the commercial appeal of Silverside Shrimp. Shrimp growers may assist the industry stay sustainable and profitable by practicing responsible farming, obtaining certifications, interacting with communities, promoting research and innovation, and educating consumers.

Balancing these elements will ensure the success of Silverside Shrimp cultivation while also benefiting the seafood industry as a whole.